Science the Revealed Religion

UNVEILING THE EXISTENCE OF AN

UNKNOWN DIVINE REVELATION

IN THE STRUCTURE OF THE

UNIVERSE, BASED ON

SCIENTIFIC

PRINCIPLES

Willis Reed

Science the Revealed Religion
Copyright 2014 by Willis Reed
All rights reserved
ISBN: 1499243324
ISBN: 139781499243321
Library of Congress Control Number: 2014910049
Createspace Independent Publishing Platform
North Charleston, North Carolina
This includes the right to reproduce any portion
Of this book in any form.
Third Revised Edition Published by
Willis Reed June 22, 2019

Contents

	Page
Preface	6
Introduction	9

Questions and Answers

1. Where is God?   13

2. What is our Soul and where does it go when we die?   13

3. Stephen Hawking, renowned physicist and cosmologist supported the following as an unanswerable question. "If God created the universe, who created God."?   16

4. How does the amount of space God created relate to the human population?   16

5. How can God listen to ten million people praying at one time?   18

6. Is there any evidence that Theology supports God being in a timeless location?   18

7. Are the atoms in our body indestructible?   19

8. Is there any scientific alternative to death as as described in Question 2?   20

  Page

9. Is there mathematical proof that space and   23
   time are abstract?

10. Does the Big Rip make our existence   24
    meaningless?

11. Can we describe what Heaven is like?   26

12. Can we assume our immortal consciousness   27
    existed before birth?

13. Is mathematics always a revelation?   27

14. How does the subconscious part of our mind   29
    relate to our consciousness?

15. In her book, The Hiding Place, God found   33
    Corrie Ten Boom's sister in the deep pit of
    Hell of a Nazi Concentration Camp.
    What is the scientific explanation of this
    extraordinary experience?

16. How can we understand the Twin Theory   34
    Paradox of Einstein's Relativity when it is
    opposed to common sense?

17. Are the waves of probability in   36
    Biocentrism a possible reality?

18. Is Atheism a problem in the United   38
    States and Europe?

	Page
19. Can we describe the source of power of the intellectual giant in our subconscious mind described in Question 14?	39
20. What behaviors continuously change the quality of our existence?	43
21. What is the meaning of humankind's existence?	44
Epilogue	41
About the Author	45

Preface

Since this book involves a new concept in religion, I will outline some of things I believe in.

The entire cosmos was created from nothing. This includes the earth, our sun, and three hundred billion other stars in our galaxy. There are three hundred billion galaxies in our visible universe. Our visible universe is only 4 percent of the total universe. There are an infinite number of universes in the cosmos. Couple this with the micro world and smaller, and you describe everything that exists.

A creation always has a creator. The infinite cosmos is so complex that only God could create it. God has a divine plan for humankind. God also has a plan for me. I cannot change it. Since God created this plan, it is absolute perfection, and I am comfortable with it.

The following are some of the things I believe about the Bible. I find the Bible very difficult to understand. For example, the following are descriptions of pieces of scripture that I find the most difficult to understand:

> 1. King David made arrangements with one of his generals to have Bathsheba's husband killed in battle so that he could sleep with her. God continued to support King David and made him a successful king. When King David grew old, he had a guilty conscience and asked God to forgive him.
> 2. The following is a description of Jesus in Revelation:
> He was clothed in a garment down to his feet and girded about the paps with a golden girdle. His hair was as white as white wool and his eyes were as flames of fire. His feet were as fine brass

burning in a furnace, and his voice as the sound of many waves, and out of his mouth went a sharp two-edged sword.
3. God had a Hebrew executed for gathering wood on the Sabbath.
4. God ordered a Hebrew general to destroy a city and kill all men, women, children, and animals. Because the general killed the men only, God destroyed his family.

All of the aforementioned is obviously not divine inspiration. It is part of a portion of the Bible that is not the Word of God. I asked people who claimed to understand and believe the Bible cover to cover to explain the meaning of these scriptures, and none of them could answer me.

Most of the Bible is just too difficult for average people to understand. The small portion they do understand gives them hope for salvation or its equivalent, everlasting life. It helps them cope with the fear of death. Fear is a useful emotion, but it can sometimes overwhelm us. We all share the fear of death. One of the many unique things about this book is that it can reduce the fear of death to a comfortable level.

There is a way to understand most of God's wisdom in the Bible. You simply read the Hebrew Bible. Hebrew is a very easy language to learn. Hebrew words have multiple meanings. You can develop skills to find underlying levels of meaning. This is not a solution for everyone because it involves a lifetime endeavor.

A simple shortcut is to read the writings of rabbis of the Jewish faith. Rabbi Daniel Lapin has one of the greatest minds in the religious world. He is famous for saying, "Everyone needs a rabbi." His writings are treasures of God's word. An example of this is *Thought Tools*, two volumes covering ancient Jewish wisdom.

This book is unique for many reasons. The three major reasons are as follows:

1. It reveals an existing unknown method God chose to communicate with humankind, based on scientific principles.

2. It provides enormous evidence supporting the existence of divine creation.

3. It also contains a recommendation about how science can revitalize the enormous decline in religion.

Introduction

After studying many branches of science, especially cosmology, quantum mechanics, and intelligent design, it became obvious that science is a revealed religion. Since this book is titled *Science, the Revealed Religion*, we must have a discussion about revelation. Revelation, when applied to religion, means something communicated immediately from God to man.

The Jews say that their word of God was given to Moses face to face. The Christians say that their word of God is inspired inspiration. The Muslims say that an angel brought their word of God, the Koran, from heaven.

Judaism, Christianity, and Islam are all revealed religions. They all have different forms of revelation. Pure science suggests that our existence is meaningless. The purpose of this book is to suggest that science is a revealed religion.

Now let us discuss why science is a revealed religion. It is based on accepting the premise that God created mathematics. If mathematics tells us that a theory is true, it follows that God revealed to us that it is true. Mathematics is not the word of God, but it certainly reveals a message from God to man.

Let us now look at an example of a theory that God revealed to be true. Cosmology was basically a hypothetical branch of science until Father George Lemaitre proposed the Big Bang theory. The Big Bang theory is based on mathematics, scientific observations, and scientific experiments. God revealed to us through mathematics that this theory is true.

In biology the single cell gives us proof that God created mathematics. The single cell has DNA. What is DNA? It is simply a list of building materials. If we were building a house and given a list of building materials such as wood, pipes, shingles, nails, etc., and then asked ten people to build a house, they would all be different.

The single cell has a body intelligence that tells it to make a dog, cat, heart, lung, etc. Nobody knows where this intelligence is in the cell. It is also a powerful computer. It is more powerful than any man-made computer. The mathematics to create the software in the single-cell computer is the same mathematics used in man-made computers. We can assume that God created the mathematics in the single-cell computer and somehow inspired humans to use this mathematics in their computers.

The following is definite proof that God created mathematics. It involves the weak anthropic principle (WAP) and the strong anthropic principle (SAP).

The weak anthropic principle tells us:
> The universe's four forces and thirty-two mathematical constants are perfectly set up for the existence of atoms, atomic interaction, planets, liquid water, and life. Tweak any one of them, and life could not exist.

The strong anthropic principle also tells us:
> The universe must have those properties that allow life to develop within it because it was obviously designed.

The following is a deductive-reasoning analysis proving God created mathematics:

1. The Big Bang is a reality and proves that the universe was created from nothing.
2. Since the universe was created, there had to be a creator.
3. The universe is so complex that only God could create it.
4. Since the universe exists, the four forces and thirty-two mathematical constants were created and perfectly set up.

5. If God created the mathematical constants, It follows that God created the mathematics.

    Albert Einstein believed in a Cosmos Religion, which knows no dogma and no God conceived in man's image. He did not believe in a personal God. Neither did he believe the individual survives the death of his body.

    This book more or less supports a Cosmic Religion but not in it's entirety as Einstein outlined it in his article in the New York Times Magazine on November 9, 1930. It involves Einstein's belief that feeble souls harbor a belief in an afterlife thru fear or ridiculous egotisms.

    There is certainly an enormous amount of room for discussion about the afterlife. The following is not just a theory of an afterlife as proposed by Einstein, it has a strong foundation in the absolute truth.

    There is only one absolute truth in science. It is the absolute truth does not exist in science. However, if Scientific Laws or Principles support a situation with a structure of strong scientific logic, it can be as close the absolute truth as possible. As an example, in Question 2 of this book, the Law of Conservation of Energy strongly supports the existence of an afterlife. The conclusion is the probability of an afterlife is close to 100%.

    The absolute truth is widespread throughout mathematics, however mathematics is not a science. It can also be found in God's Natural Laws the Bible and our Constitution.

    An other consideration is, it is essential we know the sources of the absolute truth as our country evolves toward a senseless subjective reality.

Science, the Revealed Religion is not a stand-alone religion; it can only supplement conventional religions. It cannot provide morality or prayer.

There are no footnotes in this book. The sources for all quotations and references are included in the text.

Questions and Answers

Question 1:
　　Where is God?

Answer:
　　Until 1927 our universe was accepted as a steady-state universe. It had always existed and would continue to exist forever. This changed in 1927 when Father George Lemaitre proposed what became known as the Big Bang theory. We now know that the universe was created 13.6 billion years ago, and it will end its existence roughly 20 billion years from now with the Big Rip.
　　To answer this question, we will use God's mathematics in the Big Bang and deductive-reasoning:

1. Since the universe was created, there had to be a creator.
2. The creator had to exist before the creation.
3. Time, space, and matter did not exist before the Big Bang.
4. God must exist in a place where time, space, and matter do not exist.
5. Without time, space, and matter, it would have to be a spiritual place.

Question 2:
　　What is the soul, and where does it go when we die?

Answer:
　　In the study of cosmology, we find that the universe is made of two things: matter and energy. Our bodies are also made of two things: matter and energy.

What is our soul made of? Of course it is energy. When a person dies, what happens to the soul?

First, let's have a brief discussion about our consciousness. Consciousness is not understood by any of the deep thinkers of the past or present. It is the biggest mystery, next to which all else pales. This observation can also be applied to our soul. It is safe to conclude that our consciousness is our soul. Its immortality is suggested by Thomas Paine's description of the consciousness and the afterlife in his book *The Age of Reason*, as follows:

> "The consciousness of existence is the only conceivable idea we have of an other life and the continuance of consciousness is immortality. The consciousness of existence of knowing we exist is not necessarily confined to the same form or the same matter even in this life. We have not in all cases the same form or in any case the same matter that composed our bodies 20 or 30 years ago and yet we are conscious of being the same person."

The law of conservation of energy states: Energy in a closed system remains constant. Energy cannot be created or destroyed. It can be transformed from one form to another.

Our consciousness is in a closed system and cannot change. We see proof of this when our consciousness is excluded from the ten-year renewal of our entire body. When death occurs energy in our entire body transforms to thermal energy (heat). This heat along with heat from the sun completes decomposition. Our consciousness remains isolated. At a point in time our consciousness (Soul) becomes mobile and enters a wormhole.

Since energy cannot be destroyed it must be transformed. During the heating dissipation the body's atoms emit red photons. This energy in the form of infrared electromagnetic energy restores the balance of energy in the universe.

Now let us consider where the soul goes when it leaves our body. In a study of thousands of cases of near-death experiences covered in a book titled *The Science of the Near-Death Experience* by Dr. Jeffery Long, there was one common experience in all cultures, all ages, male and female. They went through a lighted tunnel, which ended in an indescribable beautiful place. What they described was a wormhole. What it is like to go through a wormhole can be seen in the movie *Contact*, with Jodie Foster. Jodie went through a wormhole with flashing lights and came out into a beautiful place and met her father, who died when she was a little girl. Actually, it turned out in the movie that it was not her real father. The technical adviser for this movie was English physicist Dr. Paul Davies. His research interests are in the fields of cosmology, quantum field theory, and astrobiology.

Now the question is this: where does the soul go when it exits the wormhole? Since it is in a disembodied existence, it could go where God is, as indicated in Question 1, where mass, space, and time do not exist. But it is more likely to go to a place chosen by physicist-theologian John Polkinghorne in his book *Theology in the Context of Science*. It is described as follows:

"It is perfectly coherent to suppose that God will not allow the individual pattern to be lost at a person's death, but will preserve it within the divine memory. Such a persisting but disembodied existence would not in itself constitute the continuation of a fully personal life beyond death, because true human life is psychosomatic in its character. Its restoration would require God's further act of embodying that pattern in an act of resurrection in some new environment of God's choosing."

This new environment of God's choosing would have to be on a planet in a new universe. It would have to be a new universe because, as previously stated, our universe will end its existence with the Big Rip.

Question 3:

Stephen Hawking, renowned physicist and cosmologist supported the following as an unanswerable question. "If God created the universe, who created God."?

Answer:
As indicated in Question 1, God is in a place where time, space, and matter do not exist. We implied that this was a spiritual place. The past and future are necessarily as real as the present; therefore, all three domains must equally exist. We cannot answer the question in the form, "What was God doing before he created the Universe," because it involves time. Without time we can conclude God has always existed and will exist forever.

Question 4:
How does the amount of space that God created relate to the human population?

Answer:

How much space are we talking about? There are approximately three hundred billion stars in our galaxy. It would take 123 million years to reach the center of our galaxy in the fastest space ship available. There are approximately three hundred billion galaxies in the visible universe. The visible universe is only 4 percent of the total universe. There are an infinite number of universes in the three levels of the multiverse. Our earth is less than a speck of dust in the cosmos.

As stated in Question 1, our universe was created 13.6 billion years ago, and it will end its existence in approximately twenty billion years with the Big Rip. We can assume that God would not create a universe with a meaningless existence.

A hypothesis for having so many universes is as follows: Our universe is part of an infinite multiverse. All the good qualities of life are siphoned off thru wormholes to new universes. More details of this are shown in the death hypothesis in Question 2. All the poor qualities of life are discarded. This creates the following question: How do we select a consciousness with good qualities? This brings us to the age-old question: Why doesn't God intervene in the cruelties and sufferings in the world? It is obvious that he expects humankind to solve most of its own problems. There are numerous examples of this throughout history. This leaves us with one alternative: our involvement with cruelties and suffering is a trial to qualify us for an immortal consciousness.

Lee Smolen, of the Perimeter Institute for Theoretical Physics in Waterloo, Ontario, has another hypothesis. It is called "cosmological natural selection." The following is a brief description of CNS:

> "Collapsing black holes give birth to new universes by producing the equivalent of a Big Bang, which creates a baby universe with
>
> slightly different constants. In a process

analogous to Darwinian Natural Selection, those universes best able to produce would be expected to dominate the Universe."

There does not appear to be any reason why these two hypotheses could not coexist.

Now let's get back to how the amount of space relates to the human population. Humans (Homo sapiens) have been dying on Earth for two hundred thousand years and hopefully will continue to die for billions of years into the future. There may be an infinite number of planets with humans just like Earth in the cosmos, contributing to the death toll. The bottom line is this: it is assumed that the death rate of humans in the cosmos is growing exponentially. To avoid a "no vacancy" sign in heaven requires an infinite amount of space. It is reasonable to assume that God's plan is to stabilize the population growth as it evolves toward perfection.

Question 5:
How can God listen to one million people praying for six minutes at one time?

Answer:
It would take eleven years to hear one million people pray for six minutes. Since God is in a timeless place where the past present and future have equal existence, it would not be a problem.

Question 6:
Is there any evidence that theology supports God being in a timeless location?

Answer:

Yes, there is evidence in the writings of theologians Saint Augustine, Saint Aquinas, and Boethius. They understood God to be completely outside of time, looking down from eternity to the whole of the creative history with the space-time continuum laid before him, the divine God, all at once.

Question 7:
Are the atoms in our bodies indestructible?

Answer:
Atoms structuring Earth were created in a star at some unknown time before the creation of Earth, five billion years ago. All stars are atom factories. The huge mass of a star pressing down on its center due to intense gravity creates nuclear fusion. Nuclear fusion creates atoms throughout the life of the star. When a star dies, a great explosion (called a supernova) completes creation of all ninety-two elements occurring in nature and spreads them out into the cosmos.

Since nuclear fusion is not possible on Earth, additional atoms cannot be created. Therefore, the twenty-five elements in our bodies originated in a star factory. These atoms are the building blocks of our body cells. When a body cell deteriorates and dies, a new cell is created. The atomic building blocks must be reused because new atoms cannot be created on Earth. In consideration of the comments above, we could say that atoms are indestructible. Can we assume that if we look back in time before the creation of Homo sapiens, the same atoms in our bodies were in use in some other form? Were they the building blocks for trees, snakes, Neanderthal man, etc.?

Our body cells are completely renewed every ten years. We all have ten-year-old bodies. Due to entropy this replacement deteriorates over time, and we grow old. As stated in Question 2, our consciousness is the biggest mystery, next to which all else pales. Now let us consider another great mystery involving atoms: if we accept the premise that atoms are indestructible and can only be created by stars, how do we account for the following? When a human is created in the womb of a woman, where did the atoms come from?

We can make the following hypothesis: the only way imagined for atoms to get into a womb is for God to create them. It is obvious that God created the hydrogen atoms in the Big Bang. Granted, the temperature was high enough for nuclear fusion to make helium atoms, but to start the process, you would need the basic hydrogen atom. If God created the hydrogen atoms, it is not much of a stretch to create the more complex atoms.

The answer to the question is that we see proof all around us that atoms are created when animals and plant life come into existence. If we accept the Big Bang theory, atoms will be destroyed in the Big Rip. The answer to the question is that atoms are destructible.

Question 8:
Is there any scientific alternative to death as described in Question 2?

Answer:
Yes, a branch of science called biocentrism describes death this way: "In death we are both alive and dead outside of time in a timeless, absence of space cosmos."

This sounds wacky, and you may not want to read on, but if you make a study of biocentrism, you will find it to be a fascinating branch of science.

Now let us start by trying to understand a timeless, absence of space cosmos. We will make plans to take a trip to the center of our Galaxy. In Plan 1 we will use the fastest spaceship available. In Plan 2 we will assume that we can travel at the speed of light to get there.

Plan 1: Travel 150,000 miles per hour to the center of the galaxy in the fastest spaceship available.

Calculation result:

> Distance to the center of the galaxy = 27,000 light years.(162 quadrillion miles)
>
> Time to reach the center = 123 million years

Plan 2: Travel 671 million miles per hour (speed of light) to the center of the galaxy.

You are in a spaceship and about to start the trip to the center of the galaxy. When you blast off, you look in the telescopic rearview mirror. You see the crowd cheering and jumping up and down. As you pick up speed, the crowd's activities go into slow motion. When you hit the speed of light, the crowd's activities freeze, and time stops.

Calculations result (using the Lorentz transformation):

1. $T1 = T2 \times \sqrt{1 - v^2 \div c^2}$
$= T2 \times \sqrt{1 - (671 \times 10^6)^2 \div (671 \times 10^6)^2}$
$= T2 \times 0 = 0 = (Time\ does\ not\ exist)$

> Where:
> T1 = time on spacecraft
> T2 = time on Earth
> v = velocity of spacecraft = 671,000,000 miles/hour
> c = velocity of light = 671,000,000 miles/hour

2. Distance = time x velocity = T1 x v = (0 x 671,000,000 miles/hour) = 0 = (distance does not exist)

Conclusion:

Time and space are illusions of the mind. They do not exist in reality. What we visualize as time is a series of events. We can think about it as a phonograph record. The needle is the present; the music played is the past; the music yet to be played is the future. The past, present, and future on the phonograph record all exist in the same domain. Consequently, when we die, as previously stated, we are both dead and alive outside of time in a timeless, absence of space cosmos.

This may still seem to be in Wackyville, but consider the following: mathematics tells us that it is true. It follows that God tells us that time and space are illusions. Biocentrism has its roots in quantum mechanics. World-renowned physicist Richard Feynman made the following statement: "It is safe to say nobody understands Quantum Mechanics." Yet it is the most accurate branch of science. It is used to design all kinds of things. Most of the people who use it don't even think about understanding it. The bottom line is this powerful tool is only understood utilizing God's mathematics.

To try to understand biocentrism one must have some knowledge of quantum mechanics. World-famous biologist Robert Lanza suggests that the continued development of biocentrism from the microscopic realm into the macroscopic realm is the answer to the "unified theory of everything." The search for this theory in physics alone by Einstein and others for many decades is now hopeless.

When biocentrism is completely developed in the microscopic and macroscopic realms, it may be easier to understand.

Question 9:

Is there mathematical proof that space and time are abstract?

Answer:

Yes, consider the following science-fiction story:

Our country is facing a seemingly endless war with terrorists. Our cities are being devastated with all kinds of deadly weapons. People are migrating to safe havens, like Earth II, a planet orbiting the star Rigel Kentaurus. It is 4.2 light years (twenty-eight trillion miles) from Earth and is our closest star.

Bob is one of the astronauts assigned the job of transporting people to Earth II. After blasting off, Bob accelerated to a speed of 660 million miles per hour. (This is close to the speed of light.) Because of the great distance, he would normally use a warp engine to transport people to Earth II. For the purpose of this trip, a high rate of speed is more appropriate. Accordingly, it took four years to reach Earth II, and the distance was reduced from twenty-eight trillion to 3.6 trillion miles.

The plan was to stay one year on Earth II and return to Earth. At the end of one year, they blasted off again, and, traveling at the same speed, arrived on Earth in four years. As planned, forty-nine years have passed on Earth, and the war has ended.

The passengers were put into hibernation to solve the food and boredom problem. To give up nine years of life was not a consideration since science has extended life to an average of one thousand years. (This is not too farfetched; science can now extend the life of mice ten years by gene manipulation. In humans this equates to one thousand years.)

When they landed they were nine years older. After forty-nine years, hardly anybody remembered them. Due to the extended longevity, most of their relatives and friends were still alive but a great deal older (forty years older). However, they did succeed in being safe for forty-nine years.

This story tells us that God revealed to us through mathematics that space and time are abstract.

Question 10:
Does the Big Rip make our existence meaningless?

Answer:
In the Big Rip theory, our universe will be completely destroyed. The following is a brief description of this theory:

The universe is driven by a mysterious force called dark energy or antigravity and is flying apart. The furthest galaxies are moving ever farther from us, and the rate of expansion is accelerating. At a point in time, all galaxies beyond our own (traveling much faster than the speed of light) will have flown so far away from us that they will have become invisible. This is not a violation of special relativity; the galaxies are not moving. In an expanding universe, space itself expands. Visualize space as a loaf of raisin bread expanding in an oven, and you are sitting on a raisin, measuring the speed of another raisin. The raisin you are sitting on is traveling at the same speed as the raisin you are measuring but in the opposite direction. The actual speed of the measured raisin is reduced by one-half of the measured speed. Both raisins are not moving relative to the loaf of bread. At a point in time, our Milky Way begins to fly apart. Then the planets and their parent suns explode. This is followed by the atoms and their nuclei breaking apart.

The last part of this theory states that atoms and their nuclei break apart. It is based on the theory that the components of the atom are held together by gravity and a strong nuclear force. At a point in time, dark energy (antigravity) becomes the dominating force, overwhelming gravity and the strong nuclear force in the atom, and it breaks apart.

Even though the universe will be destroyed by the Big Rip, it has no effect on the meaningful existence of the inhabitants of Earth. That was cared for as outlined in Question 2: billions of years ago, before the end of our sun's life, when it ran out of fuel. Selected inhabitants of Earth were sent to planets in new universes, utilizing a death procedure with eternal existence.

Conclusion

It can be assumed that God would reuse the massive nebula of atom components of the Big Rip to create a new universe rather than to leave them as junk. The force of dark energy will weaken with expansion as the nebula spreads out, and the universe will stop and start collapsing at an accelerating rate. It will end in a big crunch, creating a singularity of infinite mass density. The four original forces will combine and compress toward a microscopic point of infinite energy. This force will be unstable, and at some point, it will explode into a new Big Bang. Inflation will spread the nebula into the cosmos for the birth of a new, updated universe.

The Big Rip may be one of God's ways to deal with entropy. Namely, replace worn-out universes with new ones. It is equivalent to replacing an old car with a new, improved model. There should be no gloom and doom in the coming of the Big Rip.

Question 11:
Can we describe what heaven is like?

Answer:
When we first think about describing what heaven is like, it seems absolutely ridiculous. However, if you consider some of the answers in the previous questions and science, you can come up with a rough description.

As indicated in Question 2, a person's disembodied consciousness will go through a wormhole to a planet in another universe. It has to be another universe because our universe will be destroyed in the Big Rip. When our consciousness exits the wormhole, resurrection will take place, and our consciousness will be embodied. The new bodies will have a unique appearance for identification. In addition there will be full awareness of our previous existence, and our names will remain the same. Also, there will be a higher level of intelligence and more than the five original senses. Death, procreation, sickness, pain, and entropy will not exist.

Since the planet will be orbiting a small star with an infinite life, there will be daytime and nighttime and a very mild change of seasons. The weather conditions will be excellent for the entire year.

Since animals have a consciousness, they will also be embodied. No animal will be dangerous, and they will not be used for food. Food will be abundant and delicious.

There will be higher levels of heaven. Those that qualify may go to higher levels via wormholes. The spiritual level will be the highest. Family members, friends, and pets will be grouped together.

Question 12:
Can we assume that our immortal consciousness existed before birth?

Answer:
The transition from death to eternity tells us that there are no recurring events. It follows that God created our immortal consciousness at birth. This is supported by the fact that there were 6.7 trillion living souls in the world in 2010, and God created the first Homo sapiens approximately two hundred thousand years ago. God did not simply create the first Homo sapiens. The first step was to create our universe 13.6 billion years ago with four forces and thirty-two mathematical constants perfectly set up to create life. The next step was to create Earth 4.5 billion years ago. Life began on Earth with the single cell approximately 3.5 billion years ago. As life evolved mammals came into existence, and the first primitive consciousness was created. It was a primitive consciousness and not immortal. The more advanced mammals, Homo sapiens and selected animals, were created with an immortal consciousness.

Conclusion:
The immortal consciousness does not exist before birth.

Question 13:
Is mathematics always a revelation?

Answer:
No. If the assumptions made in a theory are incorrect, the results produced by mathematics would be useless. An example of this is the assumption in Einstein's theory of relativity that the universe is homogeneous. If the universe were not homogeneous, the theory would be meaningless.

The following is a simple example of how a wrong assumption about time makes the mathematical result incorrect. The mathematics involved is as follows:

$$speed = \frac{distance}{time}$$

A pulse of light is sent from Earth to Mars. The assumption made in Newtonian physics is that time is absolute. It is also assumed that distance is not absolute. This is obvious since the earth is revolving and at the same time orbiting the sun. Our solar system is orbiting our galaxy, and our universe is expanding and accelerating above the speed of light. The observers on Earth and Mars would not agree on the distance since it is not absolute. They would agree on the time since it is absolute. They would measure different speeds of light.

The fundamental postulate of Einstein's theory of relativity is that the laws of science are the same for all freely moving observers, no matter what their speed. This idea was extended to include Maxwell's theory and the speed of light. All observers should measure the same speed of light, no matter how fast they are moving.

We can now rewrite our equation as follows:

$$speed\ of\ light = a\ constant = \frac{distance}{time}$$

If we analyze the above equation, we see that the speed of light is a constant and will not change. The ratio (distance / time) contains a distance that we established as not absolute. We can see that any change in distance would have to create a change in time. We now know that time is also not absolute.

Now if we send a light pulse from Earth to Mars, both observers will agree on how fast light travels. Each observer would have his or her own measurement of time and calculate his or her own time-related distance. We can now accept that time is not completely separate from and independent of space, but it is combined to create a concept called "space time."

Conclusion:

The aforementioned shows the importance of correct assumptions in creating theories. It also describes a simple example of how an incorrect assumption about time resulted in a wrong mathematical solution. When we became aware that the speed of light was the same for all observers—no matter what their speed—an analysis of the mathematics revealed an understanding for "space time." No one, including Einstein, could possibly understand "space time" without the associated mathematics. It follows that God revealed to us a description of "space time."

Question 14:

How does the subconscious part of our mind relate to our consciousness?

Answer:

Consciousness is the biggest mystery, next to which all else pales. This is also true for the subconscious part of our mind. In dreams we experience a consciousness almost completely isolated from our wakened consciousness. We must exert considerable effort to remember dreams, or they will fade away quickly. There lurks down in the subconscious mind an intellectual giant. The goal of meditation is to have packets of wisdom from our subconscious mind pop into our consciousness. An example of this can be found in the life of Rene Descartes. He had one of the greatest minds in history. His fields of interest were law, philosophy, theology, medicine, and science. He was always a frail individual and would usually spend most of his mornings in bed, where he did most of his thinking, and fresh from his dreams, he often had his revelations.

We know very little about our consciousness. Let us review some of the things that we think we know but that may be impossible to understand. The most current information about consciousness is Robert Lanza's new theory, biocentrism. Robert Lanza, MD is considered one of the leading scientists in the world. He is called the Einstein of biology. He has hundreds of publications and inventions and over two dozen scientific books. When he was a teenager, he altered the genetic makeup of white chickens and made them black chickens (in his basement). He is presently involved in an FDA trial to cure some forms of blindness using stem cells.

In the theory biocentrism, the consciousness has great power. The following are two examples of this:

1. Without the conscious observer, the universe would not exist or, at best, exist in an undetermined state of probability waves.
2. Without a conscious observer viewing it, the moon would not exist in reality.

If you analyzed these two examples, you would conclude that they are illogical and impossible to understand. However, biocentrism's foundation is a branch of science called quantum mechanics. World-renowned physicist Richard Feynman said that it is safe to say that no one understands quantum mechanics. Yet, the associated mathematics makes this the most nearly accurate branch of science.

The above examples (1 and 2) are based on the results of scientific experiments and observations. The following is a simplification of these experiments. But first, we must have a discussion about particles, such as photons and electrons. In the past there was a controversy about photons and electrons. Some considered them wave forms; others believed them to be particles. We now know that they can be both wave forms and particles, but not at the same time.

This is called "wave particle duality." Now, let's get back to the experiment. We generate either photons or electrons and shoot them at a screen (equivalent to a TV screen). They register on the screen as wave forms. When we make a conscious observation of the screen, the wave forms collapse, and we see bullet-like particles hitting the screen.

We must understand that we live in a "wave particle duality," universe. Our universe, including the moon, is in an undetermined state of probability waves. They do not exist without an observer. In example 2, when we look at the moon, the wave form collapses, and our consciousness creates the moon.

Conclusion

Mathematics tells us that the bizarre description of biocentrism is revealed by God to be a reality. Why did God make biocentrism impossible to understand? There can be only one answer: it is in the involvement of consciousness in biocentrism. As indicated earlier, our consciousness is our soul. The mystery of the soul is an eternal mystery. Our search to solve the mystery of the consciousness is over. Our consciousness (soul) is an eternal mystery.

As previously stated, no one understands quantum mechanics. We can also say the same for biocentrism. They both describe a reality that is beyond bizarre. The following are some examples:

1. A single particle can be in two different places at the same time.
2. A radio signal or light wave sent to the center of the galaxy, a distance of twenty-seven thousand light years (162 quadrillion miles), would take twenty-seven thousand years to get there. Twin particles separated by the same distance can communicate with each other instantaneously.
3. Also, as previously stated, without a conscious observer, our universe would not exist.

Since the mathematics in quantum mechanics accurately describes our existence, we can assume the following: the true existence of the cosmos is what we are aware of in the macro world and what we cannot possibly understand in the bizarre domain of the micro world.

Question 15:

In her book *The Hiding Place*, God found Corrie Ten Boom's sister in the deep pit of hell of a Nazi concentration camp. Is there a scientific explanation for this extraordinary experience?

Answer:

Corrie Ten Boom (a Dutch Christian) and her entire family were arrested in Nazi Germany in 1944 for helping Jews escape during the Holocaust. In the Ravensbruck concentration camp for women, Corrie's sister Betsie found God and became almost oblivious to the horrors of her surroundings. However, never strong in health, she grew steadily weaker and died on December 16, 1944. Some of her last words to Corrie were, "We must tell them what we learned here. We must tell them that there is no pit so deep that He is not deeper still. They will listen to us because we have been there." Corrie became an author and speaker and carried this message to sixty-four countries.

The following is a scientific explanation of this incredibly moving true-life experience. We have very little control of thoughts entering our consciousness. They seem to occur by themselves. One becomes aware of this with the difficulty of meditation. We cannot control the brain any more than we can control our liver, kidneys, heart, etc.

Now let us consider what happens when thinking stops and your mind takes a vacation. This happens when we make observations of our external world. As an example, it happens when we look at a tree, read a book, look at a movie, etc.

Under normal conditions, when we are nervous and thoughts are racing uncontrollably through our minds, we make an observation and take a residence where lights shine more brightly and where things seem more real. In Betsie Ten Boom's mind, her thoughts and emotions were unbearable. She faced death, starvation, lice, unsanitary conditions, long hours of hard labor, beatings, etc. When she made an observation, she took residence in the deep pit of hell. Under these conditions, God stepped in, and Betsie took residence in his presence.

Conclusion:

Most people who read the book *The Hiding Place* would have no doubt that God was in Betsie's presence. The question is why did God only help Betsie? There is no way that we can answer this question. We can only assume that Betsie's experience gave comfort to Corrie. This comfort was passed on to the other inmates, with prayer meetings twice a day. In addition, Corrie carried this message to sixty-four countries, which had an enormous impact on the world.

Question 16:

How can we understand the, "twin theory paradox," in Einstein's theory of relativity when it is opposed to common sense?

Answer:

First let us briefly describe the theory. John and Bob are twins. Bob travels to some distant location at nearly the speed of light. John remains on Earth. Eventually, Bob turns around and returns to Earth. Although he thought he was on a brief trip, he found that his twin brother, John, became an old man.

In Einstein's theory of relativity many paradoxes are predicted. As we approach the speed of light, many unbelievable things happen, as follows:

1. Time slows down, and we age at a slower rate.
2. The distance we are traveling decreases.
3. The mass of our spaceship increases.

All these things are observed in the perspective of the people on Earth. The passengers of the spacecraft have a perspective that time, aging, distance, and mass show no change.

It took decades for some scientists to think that they understood relativity. This understanding was mostly based on mathematics, observations, and experiments. The mysteries are still unbelievable and defy human logic.

Let us pause now and think of all the mysteries of the cosmos and the unimaginable power of the creator. These mysteries are a gift from God. Think of what the world would be like without them. Humans would know all that there is to know. Our intelligence would be at the level of primates and would not evolve.

However, we must continue our quest to solve mysteries. As an example, let us try to understand why excessive speed reduces the rate of aging. In doing so, let us use the new science of biocentrism. This is a theory that assumes that time does not exist. What we perceive as time is just a series of events. Consider our brain as a camera, taking a series of pictures. Let's call them frames of existence. As we increase the speed of our spacecraft, these frames of existence slow down. The end result is simply that less existence causes less entropy (aging). Of course this doesn't even put a dent in the mystery of aging, but it does provide some clarification.

Conclusion

All of the above is just an example of how little we know about the cosmos. We can only visualize 4 percent of the universe, and that is mostly a huge mystery. Add to that the existence of an infinite number of universes. Is it any wonder that cosmologists feel that they know next to nothing about their profession?

Question 17:
Are the waves of probability in biocentrism a possible reality?

Answer:
If you ask for a description of biocentrism, you will usually get the following answer: it is a branch of science that supports a notion that the moon does not exist when you are not looking at it. It is in a state of waves of probabilities without a physical existence.

It gets weirder. It also suggests that the universe was not created by the Big Bang. It was created by the observation of a human consciousness. Before mammals existed, the universe had no physical reality. It was randomly scattered atoms with waves of probabilities.

Here is another example: A woman turns out the light in the kitchen and goes to bed. The refrigerator in the kitchen no longer exists. I find this bewildering when I lie in bed at night, hearing my refrigerator make ice.

Proponents of this science say that it may be difficult to accept, but it is closer to reality than the Big Bang, which claims to create all this stuff from nothing. If you believe in creation the implication of this comparison becomes invalid.

Now let us assume that the universe was created by the Big Bang instead of the observation of waves of probability in biocentrism. We will now add a new theory that is compatible with the Big Bang and call it "the invisible universe theory."

"The Invisible Universe Theory"

In the beginning the universe began expanding from an infinitesimal volume—called a singularity—with extremely high density and temperature. The universe cooled to form protons, neutrons, and electrons. Thousands of years later, atoms formed. Giant clouds of these primordial elements later coalesced though gravity to form stars and galaxies.

The light waves from the stars on all formed matter were reflected and absorbed, creating "waves of description." The entire universe is invisible because light waves are invisible.

Humans, all animals, and birds have eyes to convert invisible light waves to shape and structure in either color or black and white. Insects only perceive light, darkness, and movement. The unobserved portion of the universe is always invisible.

Now let us observe the creation of an apple tree by visual observations: Imagine a beautiful red apple tree. The tree is reflecting and absorbing light waves from the sun. When a human observes the tree, bands of light waves about to enter his or her eyes represent an invisible apple tree. It is invisible because light waves are invisible. After the light waves enter the eyes, they are processed though rod and cone receptors, bipolar cells, ganglion cells, the optic nerve, and the visual cortex in the back of the head. The form and colors of an apple tree are created. This procedure allows a human to navigate around an invisible universe, which is overwhelmed with formed matter, emitting "waves of description." All mammals use somewhat the same procedure.

Conclusion

There is no doubt that the invisible universe theory describes the world we live in. It is supported by the Big Bang theory. The conversion of light waves to form an apple tree in the visual cortex is consistent with neuroscience. However, there is also strong evidence to support a part of biocentrism. To coexist with the invisible universe theory, we must disregard the moon, refrigerator, and universe disappearing into waves of probability.

Question 18:
Is atheism a problem in the United States and Europe?

Answer:
The number of people attending church in the United States is plummeting. Yet, a Gallup Poll (2010) indicates that 92 percent of Americans believe in God. Atheism is not a problem in the United States. There are many reasons for the low church attendance. The following are the most destructive:

1. The Catholic Church scandal. This has reduced the attendance in all churches.
2. Extreme friction between the different religions.
3. A history of religious violence. Examples: the Inquisition, the Mormon massacre, and the Salem witch trials.

Atheism is an enormous problem in Europe. Euro barometer surveys since 1970 show a sharp increase in atheism in Europe. The greatest increases are in predominately Catholic countries. Some say that God is dead in Europe.

The following are the current results of the series of surveys:

England	20% Atheists
Germany	25% Atheists
Netherlands	27% Atheists
France	33% Atheists

In many countries the birth rate is lower than the death rate. It is predicted that this will spread to all countries. Procreation continues to decline due to the effectiveness of contraceptives and the materialistic culture of atheism.

If this trend continues, Caucasian Europeans will become extinct, and Europe will become a Muslim continent. The accepted reason for the increase in atheism in Europe is the increase in the number of intellectuals. This is a new breed of intellectuals, called "PORGIs" (Proof or Refutation Generator for Intuitionistic Propositional Logic). At the present time, they dominate academia. It defies scientific logic that an intellectual can be guided by intuitive logic.

If this seemingly hopeless situation concerns you, I suggest that you read *America-Lite, How Imperial Academia Dismantled Our Culture*, by David Gelernter. David Gelernter is a professor of computer science at Yale University.

Question 19:

Can we describe the source of power of the intellectual giant in our subconscious mind outlined in Question 14.

Answer:
>
> The traditional sources of God's Word are:
>
> 1. From God, given to Moses, face to face
> 2. The Bible
> 3. The Koran

God also provides creative ideas and just plain help to individuals in their dreams, with inspirational pop-ups, from the subconscious mind to the conscious mind.

I became aware of this many decades ago while attending Cornell University. A close friend and I were working on a very difficult homework problem for a long period of time. At about 11:00 p.m., I suggested that we give up for the day, go to bed, get some rest, and continue working on the problem the next day. I was exhausted and fell asleep immediately. It seemed like I spent the whole night trying to solve the problem in a dream. When I awoke the next morning, the solution to the problem popped up into my conscious mind.

Decades later in my science studies, I was amazed to find that Rene Descartes, one of the greatest minds in history, had his creations early in the morning, fresh from his dreams.

One could hypothesize that God communicated with the authors of the Bible via the subconscious mind.

Epilogue

If you got this far in the book, you will know that religion should have a strong scientific foundation. As shown in Question 18, religion will not survive without it. Quantum mechanics, cosmology, and intelligent design provide enormous evidence, making divine creation a reality. In everyday life we see creation all around us, and we are so desensitized that we cannot recognize it. We plant a tiny apple seed and see an apple tree created, and we do not question where the tree or the apple seed came from. Our entire planet was created in a star factory.

All the galaxies in our Universe were created in the Big Bang from nothing. That is an enormous amount of creation.

At this point, let us consider one of the most mind-boggling lectures I have ever experienced. It was given by Dr. Mark Whittle, of the University of Oxford, on how the universe was created from nothing, in his course on cosmology. Obviously the entire lecture cannot be included in this book. However, the simple mathematics used at the start of his lecture is outlined as follows:

As stated earlier, our visible universe is made of two things: mass and energy. If we add up all the mass in the visible universe, we come up with a total of 10^{58} kg. If we add all of the energy (equivalent mass), we get the same magnitude, but we get a negative 10^{58} kg.

Do the math:

$$Mass\ (vu) - Energy\ (vu) = (10^{58}) - (10^{58}) = 0$$

Mathematics tells us that the universe was created from nothing and remains nothing today. This strange situation does not mean we do not exist. It is the start of Dr. Whittles analysis of how so much stuff was created from nothing.

Before we go on, let us define the meaning of divine creation as used in this book: all matter and energy that exists in the cosmos was created by God out of nothing. All the controversy about creation versus evolution, highlighted by the "Scopes Monkey Trial," defined creation in biblical terms.

Now let us have a discussion about the theory of evolution and "Darwin's Dilemma." Darwin had an active part in the "Cambrian Explosion," described as follows:

> Most of the major animal groups appeared for the first time in the Fossil Record some 545 million years ago in a relativity short period of time known as the, "Cambrian Explosion." Prior to this 10–15 million year period of time there are no fossil records to show that these multi-cell animals were a product of Evolution. We can only conclude they were created out of nothing.

Darwin thought that he could disprove the results of the "Cambrian Explosion," but he never found a solution to his dilemma. If he had accepted creation as a reality and had included it in his theory, we may have had fewer wars, and totalitarian governments would be more acceptable over the past 150 years.

Conclusion

We can conclude that both divine creation and portions of evolution are scientific realities. It is portions of evolution because man was created and did not originate from apes, and who can deny that one species of birds survived because of their ability to get food with their longer beaks. If the scientific community would accept divine creation, we could visualize a future with an overcrowded church and a theologian-physicist priest giving a sermon supported by a creative evolution theory. History tells us that the theory of evolution had an enormous impact on the growth of atheism. We can expect a new theory that could be called "creative evolution" would have a similar impact on the growth of religion.

Now, let us answer another difficult question.

Question 20:

What behaviors continuously change the quality of our existence?

Answer:

Morality and evil. The only thing that creates morality is the presence of God in the individual. However, atheists and a large portion of humankind have morality but do not feel God's presence because it is hidden in their subconscious mind. In addition, there are among us individuals who are oblivious to God's presence and who create widespread evil. They will walk this earth until the sun runs out of fuel, and their existence and God's plan for Earth will end.

Without morality, the exponential growth of technology will doom humankind. Evil will always exist. We can only control it.

Consider the following statement by Jean Francois Revel in *The Survival of Socialism in a Post-Soviet Era*:

> The totalitarian phenomenon is not to be understood without making allowance for the thesis that some important part of every society consists of people who actively want tyranny, either to exercise it themselves or much more mysteriously to submit to it. Democracy will therefore always be at risk.

A religion with a scientific foundation, guided by God's mathematics, can methodically control evil.

Now, let us end this book with an answer to what is ultimately the most difficult question.

Question 21:

What is the meaning of humankind's existence?

Answer:

Logic tells us that the inhabitants of heaven do not play harps and smile at each other. They are selected individuals who chose to lead a life of morality. In addition, they are immortals, evolving toward eternity under the guidance of God. At some period during this evolvement, God will reveal the reason for humankind's existence. We must just sit back and enjoy the trip to eternity. The answer will be revealed to us when we are capable of understanding it. During this period, our worship of God will be more meaningful.

About the Author

Willis Reed attended Syracuse University as a student in the College Training Detachment (CTD) of the Army Air Corp during World War II. He participated in thirty-three combat missions over Germany with the Eighth Air Force in England.

He attended Cornell University in a four-year company program while employed at Verizon as a planning engineer. It involved undergraduate and graduate courses in electrical engineering.

After decades of study, he acquired a general knowledge of mathematics, cosmology, physics (theoretical and particle), quantum mechanics, super string theory, biology, biocentrism, and intelligent design.

www.ingramcontent.com/pod-product-compliance
Lightning Source LLC
Chambersburg PA
CBHW020713180526
4516 3CB00008B/3064